> Bill,
> It's a joy to be dynamically linked and I feel the presence of your energy. Live dynamically!
> Barry

Focalizing Dynamic Links

A Human Technology for
Collectively Engaging Source Energy
& Creating a Better Future

by

Michael Picucci

With

Barry Lipscomb

If you purchased this book without a cover you should be aware that this book is stolen property. It was reported as "unsold and destroyed" to the publisher and neither the author nor the publisher has received any payment for this "stripped book."

FOCALIZING DYNAMIC LINKS
Copyright © 2013 Michael Picucci and Barry Lipscomb. All rights reserved, including the right to reproduce this book, or portions thereof, in any form. No part of this text may be reproduced, transmitted, downloaded, decompiled, reverse engineered, or stored in or introduced into any information storage and retrieval system, in any form or by any means, whether electronic or mechanical without the express written permission of the author. The scanning, uploading, and distribution of this book via the Internet or via any other means without the permission of the publisher is illegal and punishable by law. Please purchase only authorized electronic editions and do not participate in or encourage electronic piracy of copyrighted materials.

The publisher does not have any control over and does not assume any responsibility for author or third-party websites or their content.

Cover art and design by
Cover Designed by Telemachus Press, LLC

Cover Art:
Focalizing Dynamic Links graphic on the cover and the color version included in the text designed by Leah Myers, Earth Heart Art & Design
Copyright © Leah Myers, Earth Heart Art & Design

Focalizing Graphics by Michael Fiorentino with additional design by Leah Myers
Copyright © Michael Fiorentino and Leah Myers

Author photos by Nilsa Delacruz

Edited by Anne MacNaughton

Published by Press, LLC
http://www.telemachuspress.com

Visit the author website:
http://www.focalizing.com
http://www.BusinessFocalizing.com

ISBN: 978-1-939927-26-2 (eBook)
ISBN: 978-1-939927-80-4 (paperback)

Version 2013.10.30

Printed in the United States of America

10 9 8 7 6 5 4 3 2 1

TABLE OF CONTENTS

1. INTRODUCTION:
   Focalizing Dynamic Links Optimizes Everything!   1
2. WHERE WE ARE AND HOW WE GOT HERE   7
3. WHAT IT IS   17
   - Seven Conditions for Focalizing a Dynamic Link   22
   - Destructive and Constructive Conditioned Thinking   25
   - Structural Components: a 9-Point Star   29
   - Energy Arc   37
   - Holding On   37
   - Effects of Focalizing   38
   - The Takeaways   40
4. HOW IT'S DONE   41
   - Focalizer   41
   - A Diagram: Three Principles   43
   - Focalizing An Effective Business Meeting   47
   - Grounding Agreements   50
5. AN ACCOUNT OF FOCALIZING
   A DYNAMIC LINK   53
6. A PARTICIPANT'S EXPERIENCE   59
7. DYNAMIC LINKS MAY RISE
   ORGANICALLY   61
8. CONCLUSION   69
   ACKNOWLEDGEMENTS   73
   FOR FURTHER READING   75
   ABOUT THE AUTHORS   77

This book is dedicated to

Peter A. Levine

who taught me how transformation occurs
through the melding of opposite energies,

and

Otto Scharmer

who demonstrated that there is a better future
and how we can let it emerge.

Focalizing Dynamic Links

A Human Technology for
Collectively Engaging Source Energy
& Creating a Better Future

Michael Picucci

With

Barry Lipscomb

Introduction
Focalizing Dynamic Links Optimizes Everything!

SOME OF US remember President John F. Kennedy's declaration in 1962 that we would put a man on the moon within the decade. Unthinkable at the time, yet we did it, and within that timeframe. The smartphone in your pocket is one result of that inspired achievement. A more recent example of such evolutionary change is the two-term election of an African-American President of the United States, inconceivable ten years before. Today many countries and some of the United States allow the legal marriage of same-sex couples. Such recent events hint at exponential possibilities ahead.

Dynamic Linking optimizes everything. In so doing it creates the conditions for spontaneous remission of that which ails us. It also effectively creates happier lives. Just as are my own cancers, addictions, heart disease and neuroses in remission, so also is the same possible for all kinds of personal and cultural challenges.

Source Energy—that living pulse we all share—optimizes life and beyond through the formation of Dynamic Links. An example of such a Link is the internal experience I share with Barry Lipscomb, my co-author. We have lived the Dynamic Linking experience from the inside out and are deeply inspired to share this expansive experience with others. That inspiration birthed this book. Barry and I are Dynamically Linked, and such Linking is available for all people to experience.

The term Dynamic Link came from a conversation I had with my life-partner Elias Guerrero while I was working on the last chapter of *Ritual as Resource: Energy for Vibrant Living*. I was struggling to give the terms *networking* and *community* a new and higher connotation—something I was already experiencing but without a framework through which to perceive it. After asking me a few questions about my new experiences and perceptions, and particularly about how these *felt*, Elias shot back at me, in his scientific style, "You're talking about dynamic linking—links filled with dynamism!" I was particularly attracted to the meaning of "dynamism": the quality of being characterized by activity and progress, dynamic and positive in attitude. Relating to others through Dynamic Links becomes an impetus for change.

I knew he had nailed it, and Dynamic Linking became the title of the last chapter in *Ritual*. My ability to articulate the essence of the experience wasn't fully developed at that time, but it was alive and bubbling up inside me, coming from just having ritualized a circle of twelve respected peers to be resources to me in that writing. Though not fully explainable then, there was a strong inner pull to anchor the term, the perception and the experience.

Along came Barry, a new client in my practice. He had been sucked up in one of our many human vortices in order to escape the pain of being disconnected from Source Energy. He was fragile, smart, earnest, and determined not to let an addiction kill his life's potential. His innate, immediate response to focalizing awed me.

In our sessions together we consciously allowed Source Energy to guide our process and outcome. The experience was new and unusual for me. The co-conscious part was assisted by the fact that, challenges aside, Barry has been a life-long spiritual investigator and seeker. His past exposures quickly vibrated in and through our focalizing, demonstrating the most remarkable results I had seen till then. And believe me, I had already seen hundreds of miracles, many previously inconceivable. Because he quickly became a client who not only studied my work but also added fresh ideas, he soon became a colleague.

Barry compressed and reorganized my findings from a perspective more holistic than was available to me. I had been focused more on my own newest learning and teaching, unable to see the larger picture. From this grander view and our enjoyable, dynamic experience in planning workshops, I spontaneously invited him to write this book with me. I knew we could write it from within the Dynamic Link experience itself.

He replied affirmatively to my invitation, and this little book is what has emerged. Barry and I knew we shared an ongoing Dynamic Link and consciously rejoiced in it. Our collaborations were effortless!

Though the choice of my writing in the first person is for ease, Barry's energy and stamp of approval is nevertheless on every word, many of them his own. We also both share such Links, to varying degrees, with other individuals and organizations.

In the year 2000, I received the Outstanding Leadership in Research Award from the National Institutes of Health, National Institute on Drug Abuse and the National Association of Professional Addiction Counselors. At the luncheon where I received this notable award, a senior NIH researcher sat with me and we became deeply engaged in discussing my research.

I was honored that he was fluent in the nooks and crannies of what I had discovered and written about. As our lunch ended he took my hand in a fatherly way and said, "Michael, as eye-opening as your findings are, you may be in for a disappointment. Change is glacial. It will take twenty years for mass acceptance of this, and that's only if you are fortunate enough to build demonstrable models." I spurned that warning then, thinking that people would certainly gravitate to what had already proven itself to me and to hundreds of research participants. Now, in 2013, I hear the wisdom in his kind words.

Focalizing and Dynamic Linking are the demonstrable models. They can be available to every human being as the technology grows organically, and can, along with other practices, create a remission of our collective insanity (doing the same thing, expecting different results). This human technology gracefully offers innovation in our actions, allowing fresh results to occur naturally.

Renowned cultural anthropologist Joseph Campbell suggested in his PBS *Power of Myth* series with Bill Moyers that "until we center *ourselves in ourselves*, we will remain adrift, without the gift of our own inner-wisdom." I believe that is happening now. If you have not read *Focalizing Source Energy: Looking Within to Move Beyond*, then you may want to do so as it may help you to more fully comprehend the energies the authors bring to *Dynamic Linking*.

Barry and I conducted weekly out-of-state phone calls—every one uplifting—as we tapped into Source Energy to co-create a human technology capable of curing our ills—seen, unseen, personal or global.

This book does that by first presenting the context for this new technology before describing what it is. The steps to take to incubate, cultivate and nurture a Dynamic Link are then presented. They are followed by a guide to focalizing an effective business meeting as a practical application, an account of the focalization of an actual Dynamic Link, and the experience of a participant in that session. Finally there is a report on an informal Dynamic Link that occurred organically and spontaneously.

Don't let the small size of this book fool you. If you read on in an embodied way, taking a few deep breaths, giving your body permission to relax and to slow down just a bit, you will be profoundly affected. This book will guide you to a happier future.

Please share this journey with us. Allow for surprises and questions and allow them to mysteriously guide you. That's what it's all about! Just watch what happens.

Where We Are
and
How We Got Here

Allow me to flow through you unrestricted, and you will see the greatest magic you have ever seen.

–Klaus Joehle

JOSEPH CAMPBELL EXPLAINED just prior to his death that mythology is what provides a society with a reflection of its plight, grounded in the present and evincing a sense of security or stability in its common journey. He also suggests that things have changed so quickly in the past fifty years that we have been unable to create new mythologies to mirror our current existence and give us a sense of stability in the present. Our old mythologies and rituals, be they cultural or religious, simply fail to meet our current needs.

They have become outdated. We are changing so quickly that there is no time to develop new mythology to give perspective to our experience, so we are left stranded and feeling isolated from the world around us. Myths used to give us a context and a stability, to explain our experience, to put it into perspective, thereby holding the human consciousness intact and encouraging us to act with responsibility, respect and concern for the world around us. Absent myth, this stability has dissolved and we need new modes of being that will help us consciously bind ourselves to one another and to a common purpose.

Campbell suggests that the next logical, evolutionary mythology will involve knowing one's self in a much deeper way, going within to find greater identity with others and even with our planet as a whole. *Until we center ourselves, in ourselves, we will remain adrift.* In this center of ourselves we find body wisdoms of both worlds—the rational and non-rational, and we learn the language of the inner self and are awakened to our own inner knowing. This emphasizes that we must continue to allow the opening of ourselves so that we can serve as role-models to each other in the co-creation of a "partnership" future respectfully lived from the inside out.

Campbell predicted that new mythologies emerging in the 21st century would involve "Knowing one's self." This unleashes new energies. The knowing of the self will bring us the peace, equanimity and stability that mythologies of the past once provided.

William James, father of Pragmatism, teaches us that in an ever-changing universe certain guidelines must be applied. New technologies must come out of, yet honor, old realities. James suggests that we need to lean on older working systems while discovering new facts that we can grasp and use, so as to update and integrate this link from the past with the future. New wisdoms and modalities are the products of new experiences and old methods combined, mutually modifying one another.

We need new ideas (from our new experiences) that we can ride on, ideas that will actually carry us somewhere. Ideas that are alive with possibilities and grounded in experience, not just in the abstract. When we are in the midst of such ideas, James reports, we actually "feel the total push and pull of the cosmos." A feeling runs through us, emanating from our core, that we are on the right path. And then we must test our ideas in the world of reality.

> *I think of addiction as the sacred disease. Very probably, God created alcoholism in order to create AA, and thereby spearhead the community movement which is going to be the salvation not only of alcoholics and addicts, but of us all.*
> –Dr. M. Scott Peck,
> *Further Along the Road Less Traveled*

Alcoholics Anonymous is one of the few early examples of just such new wisdoms and modalities. This community-based healing movement is a demonstration of conscious connection with others for a greater and commonly held purpose. Alcoholics Anonymous has held together for the better part of a century, without contracts or money being exchanged and with complete financial transparency. It is a system of communication and an exchange of energy where the focus is on the suffering and healing of the alcoholic rather than on personal stories or egos. In this way it is a shift from an individual, ego-bound perspective to a broader, larger ecosystemic view of the entire community.

Many participants in an AA meeting experience an expansive sensation, because they are coming there to feel who they are and how we are all connected. There is a spirited component that serves one's self and that of fellow alcoholics. Old-timers (often invited sponsors) connect with newcomers and share their own experience in recovery without monetary compensation, in ways that become a grounding resource for them personally while also supporting the budding recovery of the newcomer. It's a win-win-win, the 3rd win being the collective's. People come into AA, some choosing to remain there for many years, and others shifting under their new ground of being. The program is very fluid and one remains engaged as long as it seems to serve the present moment.

There is a *human technology* underneath AA meetings and other such non-hierarchical organizations that enables them to operate organically and—in the case of AA—has allowed it to thrive and grow into a world movement without the trappings of structure and organization. When people come together in this way they connect with something greater than themselves as a resource for their recovery. This *something greater* has the potential to connect us with the world around us more deeply than mythologies of the past ever could. In this way, AA is more than merely a community movement and is perhaps, as Peck suggests, really a model to positively affect the world as we explore other applications and realize the full potential of this human technology for our collective well-being.

It is a fact that people are 99.999 percent space, being made up of atoms, molecules, and subatomic particles that are mostly empty.

–Peter Russell,
From Science to God

In *The Global Brain Awakens*, renowned physicist and futurist Peter Russell describes the actual possibility of our collective consciousness growing into a nervous system—a one-mind of all minds, if you will. He demonstrates the real possibility of global illumination being as imminent as the threat of mass annihilation. He does this with the legitimacy of real science, substantiated by research, presenting a view of the earth as a living being with each person upon it a cell in the planetary nervous system. In this view, he sees humanity's potential as a fully conscious organism in an awakening universe.

This point of view builds on the work of another of his books, *From Science to God*. Here, Russell presents the science behind the quote above, demonstrating how everything observable in our world, even our bodies, is actually energy moving in unimaginable vastness. As dense as our world might appear, when one looks at the subatomic level there is immense space. From this perspective it is much easier to see how our inner and outer worlds are connected, rather than in conflict. Russell suggests that light energy is the bridge between these two worlds and encourages us to cross that bridge to find new meaning and deeper significance in our existence.

Reading these two books, I felt a resonance with the potential that Russell shared. It was a very electrifying experience, in contrast to how saddened I otherwise felt by the direction of everything happening around me and in the world. These books showed me a possibility that something flourishing could emerge. Imagining Russell's "Global Brain," I experienced people communicating without the need for words, relating in a graceful way with the energy of love as the lubricant in the relationship. And while that vision may seem to be more in the imagined world than the physical, it is as real, or even more so, than is my experience in the physical world. This completely shifted my experience of being human so that everything now has a different meaning, with the very real possibility of our connecting for the greater good. I realized that if I could experience this potential, I could live from there, *body and bones*. This brings Peck's insight about the potential for the community movement into the realm of real possibility.

> *Transformation is the process of changing something in relation to its polar opposite*
> –Peter A. Levine,
> *Waking the Tiger*

Peter Levine's Somatic Experiencing® is a means of resolving trauma. It offers re-regulation of the nervous system by bringing resource energy (calm, pleasure) to a less pleasant, contracted energy from an earlier, traumatic experience. When these two energies are brought together in the body, just touching at their edges, there is a release of the contracted energy—true transformation occurs. One experiences a sense of wholeness in the power of that transformation—in recognizing both energies within the same self—and then in allowing for the possibility of something greater to emerge from that otherwise unpleasant union.

In much the same way, Russell's expanded view of our wholeness as a new way of being creates the opportunity for effecting collective transformation.

> *By letting go, you allow something truly new to emerge.*
> —Dr. Otto Scharmer

Otto Scharmer, senior lecturer at MIT and founding chair of the Presencing Institute, provides a means for moving forward, with "Theory U," allowing for—as he describes it—a future that wants to emerge. Along with Presencing, Theory U is a process that shifts the inner space from which we operate, allowing for what Dr. Scharmer calls the "self and Self" to communicate with one another. This small *s* self is the experience of our individual being, and the large *S* Self is all the resources of our collective experience—or Source Energy. By suspending our internal voices of judgment, cynicism and fear, we open our minds, our wills and our hearts, allowing for energetic communication to emerge.

Scharmer has developed, through his research and the work of the Presencing Institute, a positive view for moving forward. His concept of Society 4.0 suggests the need for what he describes as Awareness-Based Collective action (ABC), or the capacity for a system to see itself objectively, sense what wants to emerge and explore this future by "doing" from within the link with this larger view. It is the ability to go within a system and then look back with an awareness of the whole. That requires that we "stay in the moment" of our present experience—which is alive, vital and all-important to our emerging future—while still envisioning new possibilities. This is challenging, yet it serves to stretch our capacity for imagination, which is so vital to co-creation.

Scharmer goes on to say that we need new *human technologies* that affect ABC as a new way of being. Dynamic Linking is introduced here as one such human technology, where *co-creation* is the conception of new idea[s] by two or more people. It affirms their relationship to and respect for an energetic universe, and includes a non-visible partner—Source Energy—in that creation.

What It Is

"SOURCE ENERGY" REFERS to the life force emanating through and connecting to all things in our universe. Every one of us is a manifestation of this energy. Becoming more aware of Source Energy and working with it can optimize our lives beyond imagination.

"Focalizing," here denotes a process synthesized from a myriad of techniques that I've learned for doing just that. In my earlier book, *Focalizing Source Energy*, I describe my experience with this force as healing and transformative. *Focalizing Dynamic Links* is an extension of that work, from the individual to the collective, from a healing modality to a human technology for co-creation through shared sensing and inspired collective action. A Dynamic Link is a shared experience of Source Energy, a dynamism.

More specifically Dynamic Linking is a way of connecting with and relating to people in order to access both inner and outer resources. It happens when two or more people come together and become a unified, entity. In this more expansive place there is a sense of flow, informed by Source Energy—the intelligence of a future wanting to emerge. Dynamic Linking occurs when two or more people share a coherent, resonant state, each interpreting the same stream of information directly from Source.

Imagine being in a photographer's darkroom with several other people, all watching while a photo is being developed. As the image begins to be visible on the photographic paper, one person may notice a tree, while another may see the girl sitting beneath it, and a third notice the puppy running off to the right of the image. It's still one photograph, yet each person is tuning in to a specific portion of the greater image. In this way Dynamic Linking provides a bigger picture of "now" along with points of projection that can help participants to sense the way things might be. These can provide a way to more forward, link the current moment to some future possibility and enable us to see the present in a clearer light.

This phenomenon is already present, as the authors have experienced. The intention for this book is to give the phenomenon a name and explain it in such a way that it can be deliberately and actively engaged, enabling us to move forward more consciously and with deeper awareness.

Buddhists speak of Indra's net as a metaphor for our interconnectedness. The image is of an extended web with a multi-faceted jewel at each intersection. There is an infinite reflection of all other jewels in each individual jewel. In this way we are each linked with one another—even if by varying degrees of physical separation. While such linking is passive and even unconscious to us most of the time, Dynamic Linking brings a conscious intention to such a relationship so that the energy shared and exchanged is for the good of each person, and for the good of us all.

It's this construct of mutual best interest for all that enables a free flow of information and creates a win-win-win resolution. Thus there is no hierarchy necessary in a Dynamic Link unless in a particular circumstance it makes sense for there to be such a structure. Even then, any structure would likely emerge organically from within a Link rather than be imposed upon it by outside forces. Any beneficial structure would serve the intention of the Dynamic Link and be energetic in nature so that it is actually *felt* by those in the Link.

Further, there is no demand or expectation of a prolonged relationship or contract in a Dynamic Link, so that everyone feels free to go in whatever direction he or she is called. That freedom allows those in the Link to access Source Energy both more playfully and more deeply and significantly.

In the past, Linking may have been more common within pre-existing relationships (family, friend, organizational affiliation, etc.) where physical closeness is a factor. Yet the intention to set up a Dynamic Link is sufficient for creating a dynamically connected relationship even if there is no other social or institutional basis for that relationship.

Dynamic Linking is a new human technology for communicating, connecting and effecting. In fact, with the proliferation of social networking platforms and the ever increasing global reach of our experience, Dynamic Links don't require a physical presence or even a meeting in person. The dynamic nature of such relationships increases the bandwidth of communication to include a language beyond words—including felt sensations and other energetic experiences—so that we connect quickly when coming together, as we realize we are never really disconnected.

There are individuals with whom I have remained Dynamically Linked who were present at my first workshop more than thirty-five years ago. And there are also individuals who were involved in the past with the institute representing my research with whom I still feel Linked—even though some have now chosen other paths. I can tune into that Link and *feel* their energy in the Link still—whether I'm in their physical presence or not. The energy of connection in these Links remains a resource for my life.

Dynamic Linking can begin to shift how we think about relationships. We have been conditioned by culture to approach relationships through our minds more than through our hearts—to consider what we can get out of a relationship. Here instead, relating becomes a full-bodied, integrated and heart-centered experience. Rather than lead the process, the mind becomes a resource for applying what is already known to new information encountered in the Dynamic Link. Instead of debating, attempting to convince one another or being in disagreement, there is harmony created when individuals come together with a common respect for one another—so that communication simply builds from one to the other. This communication comes from the energy of an emerging future where there is only that.

Dynamic Linking involves no sense of ownership or obligation. One is there because he or she wants to be, and is doing something that he or she wants to do. When generally excited about what may come from the Link, one may experience a sense of anticipation, vulnerability, safety and honor—all at the same time. There is a sense of invigoration and freedom, of not being bound to anything while on a forward-moving adventure.

At the core of a Dynamic Link is a heightened (and shared) sense of the human experience.

Seven Conditions for Focalizing a Dynamic Link
There are seven conditions that when present create the environment needed to open a Dynamic Link. These are the same conditions present in focalizing Source Energy regardless of the intended outcome of the event. As presented here, these conditions relate to the particulars of focalizing a Dynamic Link and, though they are similar to those used when focalizing Source Energy in a healing modality, there is a nuanced difference.

Establishing these seven conditions dissolves any barriers among those participating and welcomes Source Energy into a Dynamic Link.

Each participant is invited to—

- start with a willingness to be *authentic;*
  [*manifesting an intention to be real—expressing one's true self rather than presenting a false front—allowing one's inner feelings and body sensations to surface and to be expressed in a container of trust*]
- gather in a *community* of two or more;
  [*when people authentically and without judgment share their struggles, joys and sorrows in a focalized format, profound experiences take place—when two or more people gather in a focalized Dynamic Link there is always the third presence, unseen but strongly felt, that of Source Energy*]
- share an *intention* for the collective experience as well as for one's self individually;
  [*when people consciously share an intention, each honoring his or her own personal intent as well, they more effectively move into alignment with each other and create a powerful, fueling energy that supports personal and collective growth*]
- have a *belief* that this intention can be realized;
  [*belief creates experience and experience shapes belief. When one bonds with others who consciously shareheart-felt beliefs, one participates in the powerful magic of creation*]

- ground as a group through *common resources* and those of nature;
  [one can gently transform barriers by inviting resourceful thoughts, feelings and sensations to arise from one's own experience or from outside sources. In so doing, the sensation of pleasure fuses with the often bristly edge of challenges, melding a palpable, grounded, corrective experience with the potential of creating, integrating or transforming into something new]
- observe the *experience* without shame or blame;
  [honor and acknowledge whatever comes up in the Dynamic Link without shame about what is being experienced and without placing blame for it having occurred]
- bring respect for oneself, for others and for all present realities;
  [mutual respect for oneself, others in the group and for the reality of what is currently being experienced doesn't necessarily mean one condones everything, but that it is acknowledged and accepted as is, with respect for its presence in reality at this point in time]

Destructive and Constructive Conditioned Thinking

The success of an intervention depends on the interior condition of the intervener.

—William O'Brien, former CEO,
Hanover Insurance Company

This quote is often referenced by Dr. Otto Scharmer, senior lecturer at MIT and founder of the Presencing Institute and Theory U, to describe a blind spot—the inner place or state of awareness—from which most of us operate. As with Theory U, the condition of our awareness is an important element of our effectiveness in focalizing Source Energy. In other words, our interior condition, awareness or seat of consciousness—the inner place from which we operate—has a direct effect on our ability to focalize Source Energy.

In indigenous cultures the shaman or medicine-person must go through a process often referred to as the shaman's journey in order to connect with his or her inner space and effectively function in the role of an energy worker in the community. The same is true for any group member in his or her capacity to effectively focalize and hold the energetic context of a Dynamic Link.

We all struggle with negative, destructive thinking—the inner-voice dictating all the reasons we *can't* do something or feel some way. There is attention required to overcome that handicap. However, it can be as simple as letting go of the struggle, turning down the volume on that limiting voice, allowing Source Energy to inspire the integration of that aspect of one's being. To accept and love all parts of oneself, rather than to be adversarial towards any of them, leads to the changes one desires.

There is a connection between one's conditioned thinking and the quality of the inner condition. In fact, we can each look at our conditioned thinking and see how it affects that inner condition. In order to understand this connection, it is helpful to distinguish among *destructive* or *negative* conditioned thinking, *instructive* conditioned thinking, and non-conditioned *constructive* thinking.

Destructive thinking limits our sense of freedom and our connection with our most authentic self. It is in some way limiting, even obstructing, our experience and the fullness of who we are in the present moment.

Instructive thinking, on the other hand, promotes the learning of something new. It creates thought patterns that are accessible in the future when one wishes to repeat this learning or apply it to one's current experience (such as learning a new language).

Constructive thinking, in contrast to both of these, is done in the moment and goes beyond simple thinking in order to access Source Energy. It often brings a sense of knowing outside of and even beyond our own personal experience. We use the mind as a tool to solve problems and serve life, often applying knowledge gained through instructive thinking in an innovative way when we connect previously unnoticed dots through constructive thinking.

At the heart of this discussion is a reversal in the method of thinking we have been conditioned to use (which is to start with what we already know and apply that, using logic). We have shut ourselves off from the vast resources of our inner knowing—that interior place of which Dr. Scharmer speaks. This naturally limits how we look at any situation:—through the lens of our particular, personally accumulated experiences.

Here I am suggesting we reverse this by first accessing the innate intelligence of our inner knowing or Source Energy, and by then using our minds to apply the information gained to whatever it is we are currently experiencing.

In this way we gain custodianship over our conditioned thinking, with an awareness that it is just *thinking* and not necessarily our current reality. Here there is a freer flow of Source Energy and its actual embodiment, so that one is not avoiding the experience or pain (emotional or physical) in the body. It allows one to be more aware in one's body, specifically in one's heart, so that to focalize Source Energy becomes a new way of being in all aspects of our lives.

In other words, we must suspend destructive conditioned thinking in order to effectively focalize Dynamic Links. By allowing Source Energy to quietly come through the heart center and in turn the body itself, one finds the real joy of integrating, dissolving and actually sidestepping the barriers of the mind's destructive conditioned thinking.

As a healing modality, focalizing is a means to somatically (in the body) engage these barriers. One creates an organic, Source Energy-guided experience that encompasses all the conditions and circumstances in one's life along with one's inner wisdom. There is no need to revisit or even attend to the original source of the conditioned destructive thinking that is a barrier to one's wholeness. This is a rich reward of the process. In such a way, we can actually drop our "stories" and move beyond them, opening to a new mythology of living fully in the present moment.

The first step along this journey is to develop awareness—an honest assessment of which patterns of destructive thinking are limiting you. From here it is a matter of recognizing that shifting away from that pattern of thinking is actually possible, and that this realization can actually be felt throughout your body. This "felt" sense is a somatic experiencing of your inner knowing—the interior space Dr. Scharmer references. This process begins to dematerialize the conditioned patterns, thus releasing the individual, or group, and allowing for a more mutually beneficial experience. One then realizes that there is a choice of which perspective one wishes to nurture. Over time as the inner condition naturally improves one gains the capacity to more consistently recognize destructive conditioned thinking.

Structural Components: a 9-Point Star

There are nine components of focalizing, as illustrated and explained here with a nine-pointed star.

We will examine each component in the above illustration, beginning at the top point and moving clockwise around the figure.

Intention—Every Dynamic Link begins with a collective and agreed upon intention as well as personal intentions that may remain private. This is what guides the process and any outcomes. You are encouraged to make both intentions well-formed yet concise. It is best to choose something that is restricting the life force in the moment—areas of your life or work where you would like to have a more flowing experience. Intention is an internal, alive, energetic, forward-moving force directed towards desired results.

Suspension—In order to be open to being informed by a deeper intelligence you must first suspend the judging, critical voices of your mind, your personal and cultural cynicisms and your fears. As you suspend the criticism, you will move into a more tuned-in sensory experience.

Sensing—As you drop into an inner experience, you will instinctively react to internal expressions through physically felt sensation. This might manifest as a heightened sense of warmth or coolness, or a tingling. Even numbness or blocks to body awareness are to be honored as sensations ripe for transformation. Regardless of the form, sensations are indications of release and new possibility.

Imaging—When inner visuals present themselves they usually transform into yet another experience. By allowing their full expression in your mind's eye, you will enjoy their assistance in making transitions.

Resourcing—A resource is memory of a person, place or action that creates a calming sensation in our bodies. This is typically accessed by recalling a time when one felt unconditional love—from a person, beloved pet, or nature—a moment from this same day or any time from the past. We all have a wealth of such resources within of which we are often not aware. In the collective experience of a Dynamic Link we also have the resources of shared experience, with the capacity for a calming sensation across the greater body of the entire group or organization.

Engaging Opposite Energies—This is where the process can begin to feel magical. There is alchemy in merging the expansive resource energy and the otherwise contractive energy that exists in desire and the unrealized manifestation of the intention.

This alchemy allows for another possibility to emerge. That new possibility, a result of energetic mingling, will *always* offer a different perspective—and perhaps one not previously considered.

Shifts of Perception and Experience—When you have been open to the alchemy of engaging opposite energies, and allowing such wisdom to touch you, you will begin to experience subtle but powerful shifts in your perception. You might notice the removal of some barrier, a new recognition of love, or a sense of gratitude for everything in your life. The possibilities are as endless as human diversity but they will all produce the same outcome: a sense of reviving one's Source Energy and of being more aligned with nature.

Soulful Integration—As with any type of growth or expansion, there needs to be an incubation period, a time of acclimation and relaxation. Children need sleep in order to grow, and it's the same for your inner growth as well. This is the time it takes to let new insights and awareness integrate and become second nature. There is no set rule for the amount of time in this process, but however brief or long, it is deeply gratifying.

Physical Manifestation—This is the part of the process where the realization of your intention may present itself; it's the time when positive results become known in the physical world. The results may not be exactly as you imagined, but they will *always* be of benefit.

Allowing oneself to enter a focalized Dynamic Link is an expansive experience. There is a vulnerable place where we can feel exposed and fragile. This vulnerability can actually increase the bandwidth of the experience. You see, coming into the focalized Link is itself an engagement of opposite energies—our normal, daily and outer experience with the heightened sense of connection with others in the Link and with Source Energy itself. And true to the structural components explained above, this engagement of opposite energies actually provides the necessary energy and willingness to go even deeper. The more we open ourselves organically in this manner, the more we are open to receive as the bandwidth naturally expands.

A focalizing session moves one from the confines of the thinking mind into the *knowing* of the heart, recognizing the heart as the source of that knowing. This knowing of the heart is experienced through the *felt* sense. Eugene Gendlin, who coined the term in his book *Focusing*, defines the "felt sense" as "not a mental experience but a physical one. *Physical.* A bodily awareness of a situation or person or event. An internal aura that encompasses everything you feel and know about a given subject at a given time—encompasses it and communicates it to you all at once, rather than detail by detail." This felt sense contains *all* the information and one must observe it fully to uncover everything completely.

In this way, the felt sensations and new physical awareness are the gateway to inner resources and to Source Energy itself. One first experiences the innate intelligence of these resources in the heart as they then come into awareness through the sensations felt in the body. By following these sensations one can discover what Source Energy has to reveal in the situation. Peter A. Levine explains in *Waking The Tiger*, "the felt sense unifies a great deal of scattered data and gives it meaning."

While the experience of a focalizing session is quite organic and seldom unfolds in the exact order of the nine components, there is an arc of energy through the whole session that may be helpful in understanding the otherwise non-linear nature of focalizing.

Energy Arc

There are four distinct phases to that arc of energy: *Tuning, Focusing, Amplifying* and *Directing*.

Participants begin by turning their attention inward and toward an awareness of their own beings and bodies at that very moment. This is *Tuning*.

One then turns this attuned attention to the intention, and to all the multi-dimensional elements of its possibility and potential. This is *Focusing*.

That focus is then amplified by drawing in resource energy through heart-opening love. This includes the supportive energy of envisioned special people as well as the innate intelligence of the universe itself and its ability to hear the intention and to help it manifest. This is *Amplifying*.

Finally, each participant engages in the *Directing* of this energy by freeing it completely, possibly allowing it to rest in stillness, and then curiously observing as the intention mysteriously manifests itself—or even something more agreeable—in the physical world.

Holding On

Before we leave this section, it is important to note the possibility that a time may come in any Dynamic Link when the connection gets strained and doesn't feel as "juicy" as it once did. A word of caution about conditioned thinking. When relationships are not comfortable, conditioned responses may include an urge to get in there and attempt to restore an earlier, more comfortable feeling. But as mentioned above, there is no expectation of a prolonged relationship in a Dynamic Link, so everyone can feel free to go in whatever direction he or she is called. Remember that it is this freedom that invites Source Energy into the Link and allows an even better future to emerge.

So we are reminded *not* to hold on so tightly to these earlier feelings of connection when they have shifted and people go off in different directions. Instead we are encouraged to rejoice when a Dynamic Link changes, much as we did when it originally formed.

There is great freedom in being there for what you are getting in the moment; and when that is no longer there, in being able to move on—going wherever your inner voice calls you. Even so, your connections and contributions in the Dynamic Link will remain intact, if more remote. There is no sense of abandonment. These Links always exist, even if not active. Once you have shared the space of a Dynamic Link with someone you will always have a connection with that person—something alive and organic that becomes a part of your being.

Effects of Focalizing

Focalizing is a dynamic process that allows us to set aside day-to-day thoughts and feelings and access our innate intelligence. The process can benefit any person, organization or collective stuck in their current circumstances. With its ability to reconnect us with a natural resource that conveys new perspectives, focalizing often illuminates previously unseen possibilities for moving forward with grace. By learning to transform overwhelming life situations we can suspend our everyday thinking mind and access a timeless source of change. Focalizing allows us to develop a newly enlightened intelligence that becomes integrated into our very beings.

In focalizing, boundaries between past, present and future dissolve, allowing us to leave linear time behind and coaxing us into the realm of timelessness. This shift to a more timeless field happens so gradually that at first it is hardly noticed. From a natural state, shifts and movements in the body's felt senses, coupled with imagery, arise—and illumination and transformation are possible. The focalizing experience offers a sense of resilience, strength, and balance.

Dynamic Links are by their nature vibrant. The individual experience of Source Energy (where one has the feeling of observing at a deeper level, of experiencing greater clarity, ease and presence) can be extended to organizations and businesses, sports teams, non-profits or any other group working towards a common intention. There is a distinct sense of Resource, an uplifting experience, present in the Link. Those who are Dynamically Linked shift into a higher level of vibration, a less encumbered state, and operate from a place of wholeness.

In time this experience becomes familiar. It may feel flat to be *out* of any Dynamic Link once one has experienced the free-flow feeling of it, when the jaggedness of the disjointed energy outside of these Links becomes more obvious. This natural reaction will pass gracefully as your own inner compass rises to the fore. Often, new Links, even with yourself, become invigorators of spirited restoration.

The Takeaways
Here are listed five tangibles of a focalized experience that are present in a Dynamic Link and provide the participant with a feeling of being more alive. There are likely to be experiences of:
- feeling more presence in the moment,
- having some resilience, an ability to bounce back gracefully when shocked,
- having an ability to connect dots by removing blind spots and thereby gaining access to previously unseen possibilities,
- shifting your point of view and desires, and
- feeling inclusivity, unification and compassion.

While Dynamic Links, by their nature, occur organically and often without great effort, it is also possible to aid and even to amplify these links by using the powerful tools of focalizing. One can focalize a Dynamic Link by utilizing the principles discussed in the next chapter.

How It's Done

THERE IS A PARADOX in attempting to describe a focalized Dynamic Link; we are constrained by the limited ability of words to convey the nature of this experience. Any words shared here do not mean now nearly as much as they will when you actually experience what I have attempted to describe.

To provide the best perspective on Dynamic Linking, this chapter includes three approaches to it: guidance for the Focalizer, a diagram of the process, and, finally, an outline of a collective (in, this case, a business meeting) using these techniques.

Focalizer
While everyone involved in focalizing a Dynamic Link is effectively a Focalizer, it is often beneficial for one or more members of the group to step forward and be empowered by the group in an active role of focalizing the

Dynamic Link. In so doing the primary Focalizer(s) can help to bring focused, deliberate and conscious attention to the group's collective intent, energy or desire. They do this by embodying many of the elements of focalizing through energetic and active engagement with both the group and the collective energy of Source being focalized.

Thus, a Focalizer:
- holds the context and creates space for the collectively-held intention of the group as well as for the diverse and personal intentions of each individual;
- helps participants to shift perceptions, offers additional perspectives and opens the group to new possibilities;
- brings resource energy in the form of caring, love and a real sense of community to the intention and to the group collective;
- connects with the organic entity that is the group as a channel for the expression of its greater, innate intelligence and wisdom;
- respectfully engages opposite energies within the group for resolution, integration and the emergence of a greater idea or vision;
- holds no point of view nor expectation of a specific outcome other than that of the aligned intentions; and

- allows his or her self to be open to and participate in the creative energy of Source in the Dynamic Link.

With an understanding of the conditions and components of a focalizing session, as discussed in the previous section, one need take only a few steps in order to focalize a Dynamic Link in time and space. To illustrate the focalizing process we will use the symbol of the circle and the triangle, also used in many healing and spiritual disciplines.

A Diagram: Three Principles

Representing feminine qualities of unity and wholeness, the outer circle symbolizes the unbroken cycle of life—death and rebirth—and Source Energy. The triangle at the center is the strongest of geometric forms. Unable to collapse unless one of its joints gives way, it points upwards, symbolizing the active, male principle and masculine power. Together these illustrate a unity of body and spirit and demonstrate a balance between structure and freedom—the unity of opposites so critical to the process of focalizing.

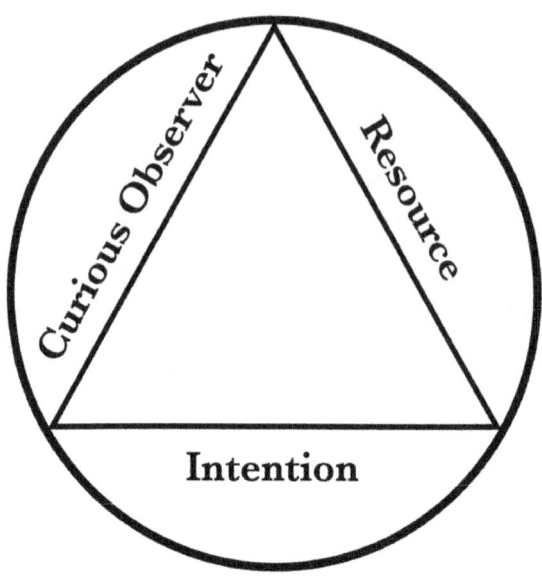

The Focalizing Structure

The three sides of the triangle represent the alive *Principles* we engage in order to focalize a Dynamic Link.

1. The base of the triangle represents *Intention*—the basis of and driving force behind any focalized Dynamic Link. Held collectively it is shared by all those in the Link. The person focalizing the Dynamic Link asks participants if they can share a *particular intention* for the group itself, and welcomes each one to also include personal intentions. The power of the Dynamic Link comes from holding together these intentions—the

collective intention for the gathering as well as personal intentions (even if those have not been shared).
2. *Resource* is then added to the intention by having participants get in touch with a shared sense of ease, joy or pleasure (Source Energy)—often an experience of accomplishment or shared emotion that can be felt throughout the group.
3. The *Curious Observer* is then brought in. That is the part of one's self wise and grounded enough to participate in the context of creation, an experience of the timeless and form-free awareness of infinity. The Focalizer encourages the group to observe what is occurring and the participants each to observe what he or she is personally experiencing throughout the shared experience of the session. This involves the capacity to simply "be," to notice, without judgment or agenda. Such awareness permits our hearts to mysteriously speak through our bodies. The Curious Observer acts as quiet host to body, mind, heart and nature simultaneously.

The diagram reminds us that the universe of Source Energy and our own innate intelligence as embodied in the Curious Observer are always present. The circle and triangle configuration illustrates the ongoing miracle of creation.

Since Resource is unlimited and universal, and the Curious Observer is much wiser than our linear-thinking brain, the outcome of a Dynamic Link may be in actuality greater than its initial intention was. The physical manifestations from such a Link may not look like anything we could have conceived of before the experience, yet they are often surprisingly agreeable, possibly more so than the original desire.

With the three sides of the triangle in place, we can use two devices to lock the energetic Dynamic Link in time and space:

- The first locking device is shared intention which the Focalizer engages by asking participants if they have an intention that all present could comfortably align themselves around., a collective intention, as well as a personal one, which can remain private. The acknowledgment of this aligned shared intention works like an arrow of energy traveling in time and space.
- The second lock is shared belief (however small) that the shared intention can actually manifest in reality. The Focalizer then affirms the group's belief in this collectively held belief and now locks the energetic and resonant Dynamic Link into place.

How I came to notice, and later trust, these locking devices was a mysterious process. They came though me when I was writing *Ritual as Resource*, as if I were channeling Source. This information arose from my inners and has survived the test of time. And of testing. I have watched these basic functions produce consistent, unimaginably good results, manifesting in thousands of courageous souls a deeper awakening of joy.

You may want to notice your own felt awareness as it whispers or sings to you. As your embodied wisdom becomes more available, curiously observe what your body has to report.

Focalizing An Effective Business Meeting

The four phases of the energy arc in a focalizing session—Tuning, Focusing, Amplifying and Directing—can serve as the outline for a business meeting and in so doing transform a typical, often ineffective, meeting into something encouraging and inspirational.

There is an awkward prerequisite for focalizing Dynamic Links in an organizational or business environment and that is that the leadership must be willing to forego typical hierarchical structure, along with its implications. It requires courageous shape-shifters, willing to link with their colleagues in a collaborative way outside of the typical organizational structure. Like other participants, they agree to be led by a future that wants to emerge through them. This place of "not-knowing" is difficult, particularly so for our leaders. Those who choose to explore this path are to be applauded.

With that vital ingredient in place, the focalized business meeting is opened by *Tuning*, creating connection and tuning into the collective energy of the organization as well as to the presence of Source Energy itself. We do this by tuning into the jointly-held organizational vision and mission (presumably already in place, and an important element of a more conscious organization).

There is an energy circulating through any group of people, even in a professional setting. When we deliberately tune in to that energy vibration (often with a simple breathing and embodiment exercise) we connect ourselves in a way that allows us to collectively access Source Energy.

It is important that we each bring our real self to the Dynamic Link. This is done by being authentic and even vulnerable. Each person takes a moment to quickly share the current challenges and triumphs in his or her functional area—honestly and fully without attempting to make themselves or their function appear any different from what is currently being experienced.

The final element of Tuning is for everyone to let go of expectations and engage an attitude of "not knowing" so that Source Energy is free to come through unfiltered and undeterred by preconceived notions or the expectations of a particular outcome.

The *Focus* of the meeting brings this vibrational energy to facilitate the matters at hand through the collective intention. It is important to be clear and to align one's focus on the intention for the meeting, however simple that may be. It is important to be true to the intent of the meeting whether it is as simple as sharing information or something more in-depth such as developing a solution for a particular situation or deciding what course of action to take with a specific issue or challenge. The collective holding, by all those present, of the shared intention is essential to moving forward in a focalized meeting.

It is also important to create space for each individual in the meeting to craft and hold their own personal intentions as well. These can be distinct and separate from collectively-held intentions; one intention need not have anything to do with another. By clarifying and giving attention to both personal and collective intentions we honor the *whole* of our lives in the process.

Next we *Amplify* the Focus to support the intention of the meeting. This is done by collectively holding the shared belief in achieving the intention, and by inviting common resources into the meeting as support. The meeting's Focalizer does this by asking whether or not everyone has a belief—even if very small—that the intention for the meeting is possible and can be manifested. After any discussion required to resolve a participant's disbelief, the Focalizer affirms that he or she too believes it is possible.

Next, the meeting's Focalizer invites everyone, through this shared belief and collective vision and mission, to *Direct* the energy of resource to illuminate the intention. Upon becoming more familiar with the group, the Focalizer can bring in more specific and perhaps emotionally-connected resources such as past accomplishments, the support of a particular person or some other collective experience. This amplifies the intention allowing the collectively held belief and shared resources to work.

Finally, the meeting's Focalizer gently guides the thus amplified Focus to engage the Curious Observer in flow with the meeting's agenda. As participants individually engage the Curious Observer, each accesses his or her own innate intelligence, welcoming Source Energy into the interplay of the meeting.

We experience our innate intelligence first subtly, in the heart, and then through sensations felt in the body. When we pursue and follow those felt sensations with the Curious Observer we are open to discovering that which our intelligence (Source Energy) has to reveal in the situation.

Grounding Agreements

The engagement of the Curious Observer is maintained throughout the meeting by honoring a few grounding agreements.

The first is that each intend to listen intently, with the full body, and attention to the rise of any sensation felt within. Participants are encouraged to listen actively to what others are saying, not be planning what to say in return. It's the deep listening that allows a response to present itself as appropriate.

The second is that each participant hold the intention that the meeting serve the higher good of every individual in the Link, as well as that of the organization as a whole.

The third is that each person operate with honesty and integrity—saying what they mean and meaning what they say—with the understanding that everyone else will do the same.

And fourth, that each agree to maintain a respect for him or herself, for others and for all existing realities (those challenges, barriers and issues that currently exist in everyone's experience).

After the business of the meeting is completed, the Focalizer concludes by having everyone honor the connection to Source Energy that ran throughout, supported the meeting's intentions, and enabled the outcomes to emerge.

An Account of Focalizing a Dynamic Link

BARRY LIPSCOMB'S ACCOUNT of focalizing a Dynamic Link in a business meeting is presented here as just one example of how such a process might be experienced from the point of view of the Focalizer.

The session opened with everyone sitting around a table. I invited everyone to simply focus on their breathing, without any effort to alter or change it in any way—merely noticing the breath as it was currently. I suggested everyone suspend any interior voices of judgment, doubt or cynicism, and acknowledged that while we can't control these thoughts' arising, we can choose not to give it attention.

Then I asked the participants to slowly deepen and lengthen each breath—drawing it more deeply into the body and allowing it to move more slowly and deliberately. As they continued to breathe, I suggested that with each breath they drop even more deeply into their bodies. After a few more breaths, I suggested they scan their bodies, and notice where they were feeling calm and relaxed, and fully tune in to that feeling.

Next I asked each participant to shift his or her attention to areas of the body where there might still be tension. I asked that they each imagine breathing directly into any place of tension, to notice how it might shift or change—even inviting some of the calm, relaxed places in the body to move into any place of stress or tension as resource energy, being aware of both feelings, and noticing the very edge where they are brought together.

They continued this for several minutes until I sensed a grounded calmness in the group—a palpable but subtle feeling—of a still and quiet place which I could sense in my own body.

Please note that this description won't fully convey the slow pace and subtle nature of the session. There is a quality of allowing the process to move at its own pace and to unfold naturally. It is also important to note that the Focalizer frequently suggests that there is no particular effort here, just an invitation to go where the experience comfortably takes one and to simply observe that experience.

Once everyone was in this space, I asked them each to shift awareness to the heart-center, the center of the *chest, while continuing to focus on the breathing. This is simply noticing the heart and inviting it into the experience in any way that feels okay. I invited them to move even deeper into the heart-center with each breath, to the core of their beings or to whatever they were experiencing, simply allowing for what felt comfortable and appropriate.*

After a few minutes, I suggested they might notice or allow a formless and timeless experience—outside normal time and space, separate from what their physical bodies were feeling. I then asked them to imagine or to allow for the possibility of each of them connecting their heart-center to the other heart-centers around the table—by string, a line, rays of light or however they might imagine and visualize the connection.

At this point I invited the consciousness, or Source Energy, of their business organization into the space all were holding. I then asked each person around the table to imagine connecting into that space through their heart-center as well.

I reminded the group of their intentions for the session, the particular areas of their business for which they sought insight and guidance. As they tuned in to these intentions, I asked that they take notice of their bodies' sensations, which could be the gateway to visual images, auditory messages or simply "knowing." They might even bypass sensation and immediately tune in to images and messages. I reminded them to be mindful and notice what came up and then share it with the group. As the first person shared what she was experiencing, it led to the next person's experience and then the next, each one building on the previous participant's.

One of the group's intentions was to gain insight into how they might expand their operation and move into a new geographic area. While the group came to the session with a strong idea of where they might locate their expansion, no one felt any sense of energy around that particular location once we were focalizing the Link.

In fact one of the participants shared a vision she had of a large shadow over that part of the map. Slowly, everyone around the table began to describe different sensory experiences—some fragrant, others visual, even sounds and feelings—as the sense of an entirely different city arose among the participants.

This was not the city they had come to the session focused upon. As everyone tuned into this different location more fully, they also began to sense a season of the year and to envision business meetings and distinct settings. Everyone reported an expansive feeling of being open, warm and generally positive, excited and optimistic.

We then moved our attention to other elements of the intention for the session, with the same experience of information unfolding as each person shared what they were sensing from a multi-sensory experience. Each of these experiences provided guidance and insight into how various aspects of the business might unfold, accompanied by a collective, positive feeling of opening and expansion.

The session also provided an opportunity for experiencing contrasting energy—something in the way, such as a barrier in one's own personal energy system, or the "thinking" mind slipping back in and creating confusion by over-thinking rather than fully feeling *the experience. As we moved through the session one of the participants experienced the integration of a personal barrier and that then cleared the way for the vision of the business to unfold more fully.*

A Participant's Experience

THIS ACCOUNT IS given by a participant in the same Dynamic Link described above by Barry Lipscomb as Focalizer for that session.

We first set our intention for the session through specific questions we had about our business where we felt we could use some higher wisdom. With those areas identified, we were guided through a grounded presence exercise.

Once we were all deeply relaxed, the Focalizer posed our questions one at a time and asked us to describe any felt sensations, images, colors, or messages we were receiving. I noticed that in the beginning of this part of the experience, we each had distinctly different insights, but as the session continued we became more and more tuned in to the same images, adding on as we went. For example once the idea of growing in Chicago surfaced, we all felt a resonance with that idea and began receiving information about the business expanding to this new location.

Also, a number of us felt that there would be a lot of forward movement in six months. We were most surprised that we all had a dark, heavy feeling when we thought about trying to move forward with our then-current plan. Not only did the Linking give us valuable, surprising and useful information, but it also brought us closer together as a team.

–Jeannie Ballew,
Co-founder of Entre-SLAM

Dynamic Links May Arise Organically

DYNAMIC LINKS NEED not be focalized in such a structured and defined setting. In fact, they may emerge organically when the conditions are right. Case in point is Jennifer Medina's experience coaching a soccer camp in Mexico.

Jen is a former client who found herself focalizing a Dynamic Link quite by chance. Jen is a soccer pro living in San Francisco. She has a passion for working with children and through a series of seemingly unrelated synchronicities she formed and coached a soccer camp in Mineral de Pozos, Mexico, a small village about 180 miles north of Mexico City.

Several events collided to create this opportunity. First was Jen's desire to learn Spanish—a small seed but a vital element to what would unfold. Second was her client Nick and his mother, Elisa. And third was an email.

Nick's mother lives in New York City but is closely connected to Mineral de Pozos, where she has a second home, built with the help of the local people, who live a simple but hard life. With a concern for her adopted village, Elisa emailed Jen with the subject line "Crazy Summer Idea."

The email went on to share this crazy idea—for Jen to travel to that Mexican village and, for four weeks during the summer of 2012, run a free soccer camp for the children there. She would also work with the community leaders to develop a long-term plan for the kids and the community. Ready for a challenge and excited by the opportunity to immerse herself into the Spanish language, Jen gladly accepted—and stepped out into the unknown.

Arriving in the village of Mineral de Pozos, Jen discovered that there had not been a lot of pre-planning. Flyers had been posted around the village, but there had not yet been any real effort to recruit kids for the program. She worked with Lincho, the town's gardener and head of children's activities, to build interest in the camp. Lincho is deeply connected to the village and seemed to always know what was going on and could make things happen. Slowly interest was spread about the program and it simply grew organically by word of mouth—with six kids the first day and then eighteen, and finally two camps by the second week.

Jen noticed that as the kids became more comfortable with one another small groups would form each day and then would organically shift and change from day to day. She remembers in particular Cesar, who naturally began to lead one of these small groups. Sensing this would be appropriate, Jen gave him responsibility for that group and his leadership blossomed.

She explains her experience as one of feeling completely in her element where there is no real sense of time and space. The environment allowed her to just have fun and make mistakes trying to speak Spanish, while encouraging the children and volunteer coaches also to feel free to try what she was teaching without worrying about failure. This openness and willingness is what enabled her to sense it would be appropriate to give this guy leadership responsibilities.

Juan (left) and Nick on a water break. You can see by their red and yellow "camisetas" (scrimmage vests) they were just playing on different teams. However they can laugh together and go on being friends during the break!

There was also a lot of sensitivity in the group. In particular there was Marco-Antonio, a deaf boy who came to the soccer program with a background of being kicked out of other programs in the past. Jen says that he was indeed a handful and his mother had never been able to get him to even try wearing his hearing aids. This shifted entirely in the group where the other kids diligently focused on including him. The shift was apparent when he brought his hearing aids to camp the second week and wore them, as his mother enthusiastically explained on the last day of camp, "for the first time ever."

Jen says another significant element of the experience was working with the women's team. These women had assumed that Jen was there for the men and boys, relegating themselves to their position beneath the men. Jen was having none of that and extended her coaching to include the women's team in the village. They even asked her to coach their championship game. Jen declined for fear of not being able to think, react, translate her thoughts into Spanish and speak quickly enough to be effective, but she assured them she would be there on the day of the game to cheer them on. Yet, the team, with a long history of disappointments and unfulfilled promises, was completely surprised when Jen was there on game day, as they had not expected her to actually show up.

Beyond these tangible yet significant outcomes, there were even deeper transformations that Jen witnessed. She says it was eye-opening to realize that because of the water quality most of these kids only drink soft drinks. This was obvious on the first day when they played in the sun for hours without ever taking a water break. Realizing their dehydrated bodies did not know they needed water, she made a point of taking deliberate water breaks and watched as they re-learned a thirst for water. In only a few days they were naturally going for water as they were thirsty. She goes on to share that this thirst for water was a metaphor of all they thirsted for—love, learning, attention and even discipline.

She says that discipline was one of the ways she brought them all together and created a bond. Jen explains that discipline is done with love, as compared to punishment that comes from fear or hate. She was genuinely concerned for each child and showed this through being non-judgmental and being in the moment with whatever was going on.

Jen received many rich rewards while in the village and says she experienced an exchange of gratitude everywhere she went. She also noticed how the soccer league created a greater sense of community as everyone came out to watch and support the team. She was especially affected when Luis, a villager in his 70s, told her with tears in his eyes that he had never seen anyone put so much love into the kids before.

All players and volunteer coaches received certificates which were incredibly important and equally meaningful to all who participated. Pictured here with their certificates are some of those players and coaches. *Top Row(left to right): Volunteer Coaches, Cesar and Manuel. Middle Row: Jose, Volunteer Coach Luis, Elisa, Jennifer, Jose. Bottom Row: Pedro, Manuel, Marco Antonio.*

Jen's experience coaching soccer in the Mexican village is a rich Dynamic Link, mutually beneficial to everyone; she still feels deeply connected with the village and each person she met there. Further it is fueled by the alchemy of focalizing when we engage opposite energies: in this case, the Latino culture of machismo amplified by the competitiveness of soccer, and Jen's being a woman. Here she was, a woman teaching the male coaches about soccer. This was transformative—for the men she worked with and perhaps more so for the girls who participated and the women who observed this.

As for Jen, the experience was empowering as she participated in the transactional nature of love and witnessed the palpable exchange back and forth. Teaching soccer in another language was a special challenge as it involved a double process of translating—first transmitting the learning and then translating that into Spanish. With her original intention to learn Spanish, she was able to relax into the moment and be present to enjoy the language while trusting that the coaching would just happen.

Learning Spanish was a backdrop to the whole experience. In the village, she met a woman who was learning English. They would get together and spend hours talking in each language and "coaching" one another. And, as evidence that Jen's experience is still deeply with her, she says her Spanish may still be broken but is stronger than many people who have spent much more time in the culture.

Conclusion

Problems cannot be solved at the same level of thinking that created them.

—Albert Einstein

IN THE LAST hundred years or so the cultures of the developed world have been anchored in linearity and duality, in black-or-white thinking. We keep doing the same things over and over again, despite the fact they may no longer be working for us. The evolution of humanity is at a turning point. The perceived "right way" to do things has been anything but.

Seeing individuals as only that, and not also as part of a larger whole, has gotten us where we are today. We need new human technologies that respectfully and effectively bridge the arc from outdated paradigms to emerging new awareness. *Focalizing Dynamic Links* has been presented as such a technology—one possibility for addressing some of the greatest challenges we face in sustaining a continued existence for ourselves and for those still to come.

Dynamic Links are characterized by the energetic engagement of constant change for our collective well-being. They offer the promise of gracing and assisting our human evolution, as more people learn to embrace and engage in this manner. With globalization, the ubiquitous Internet and the proliferation of social networking, the opportunity for forming Dynamic Links is expanding exponentially.

The next leap forward could start in many places. A dream of mine is that the United States create a cabinet-level Department of Peace. With a modest budget, such a department could bring together the world's most respected and knowledgeable peaceniks and technological visionaries to create ever-expanding prototypes of peacemaking and nonviolence for people and planet. Dynamic Linking might even inspire other countries to take similar positions, returning to the U. S. its legacy of leadership.

This vision may be of a future not so distant. All of us are gradually being met by a new world of promise and possibility. It seems we are at a point of divergence, with two choices before us. We can continue to engage with the world as we have in the past—and if we do that we are likely to realize the insanity of repeatedly getting the same poor result until a sustainable existence is beyond our reach.

Or we can take the other path, one that leads forward in such a way that we all thrive. If we chose this other way, focalizing Dynamic Links is a method for supporting it. Dynamic Linking gives "going viral" an expanded meaning and can influence those with whom we collectively engage, as well as how we do that.

It is clear we are in the middle of a crisis and that at the same time there are good possibilities available to us. Here is a vision for how our experiences could *feel* different in the future and how we might interact differently, with more responsibility and joy. Imagine a future world interconnected through cooperation, consciousness and sustainability. Seeing ourselves as more similar than different, we might realize that we are all in this together and that also our collective survival is contingent on each of us thriving individually. Recognizing our connectedness, we may find that prosperity and abundance with one another is our natural state. Rather than attempting to dominate the environment, we could see ourselves as stewards of the planet we inhabit. Rather than fight against destructive conditioned thinking, we might then each become a compassionate custodian of our individual mind, using it to craft appropriate responses to whatever we are alerted to by our physical senses.

From a focalizing perspective we give attention to new possibilities as they emerge. There is no specific "call to action," as that would suppose that we know what it is others should do. In fact I have no idea what they should do. The point of engaging in Dynamic Links is to connect with each other through the integrity and authenticity of *who we all really are,* so that we can each live from our inner conditions in the best of possible ways.

May we continue to find new ways to be alive, consciously connected to the movement of every moment, finding the foundation of new principles for stepping out into the unknown, allowing a new reality to manifest through us. With little memory of an earlier time feeling un-empowered, defeated, insignificant or disregarded—the result instead would be a thriving populace living sustainably across all measures of being—connected, contributing and conscious.

Acknowledgements

OUR GRATITUDE GOES to all who have influenced our path to the discoveries of Focalizing and Dynamic Linking. Not only our teachers named herein, and others referenced on our web sites, but to our clients through the years who have demonstrated the common sense application of these processes. The practical manifestations of goodness from these processes are our greatest inspiration.

We bow to our colleagues and friends who have supported us in trusting our own inner compasses to walk through the unknown, and be informed by it.

We honor our families and life-partners for understanding our grit to never stop discovering.

Lastly, we acknowledge Source Energy, as it is who we are, and how it lovingly guides our paths, individually and collectively. Thank you. Thank you!

For Further Reading

Michael Picucci, *Focalizing Source Energy: Looking Within to Move Beyond*, Telemachus Press

Michael Picucci, *An Introduction to Focalizing: Organic Solutions to Real-Time Challenges*

Michael Picucci, *Ritual as Resource: Energy for Vibrant Living*, North Atlantic Books

Michael Picucci, *Journey Towards Complete Recovery*, North Atlantic Books

Eugene T. Gendlin, *Focusing*, Bantam Books

Peter A. Levine, *Waking the Tiger: Healing Trauma: The Innate Capacity to Transform Overwhelming Experiences*, North Atlantic Books

Peter A. Levine, *In An Unspoken Voice: How the Body Releases Trauma and Restores Goodness*, North Atlantic Books

Lynne McTaggart, *The Bond: How to Fix Your Falling-Down World*, Atria Books

John Renesch, *The Great Growing Up: Being Responsible for Humanity's Future*, Hohm Press

Peter Russell, *The Global Brain Awakens: Our Next Evolutionary Leap*, Global Brain, Inc.

Peter Russell, *From Science to God: A Physicist's Journey into the Mystery of Consciousness*, New World Library

C. Otto Scharmer, *Theory U: Leading From The Future As It Emerges*, Berrett-Koehler Publishers

Otto Scharmer and Katrin Kaufer, *Leading From The Emerging Future: From Ego-System to Eco-System Economies*, Berrett-Koehler Publishers

www.Focalizing.com

www.BusinessFocalizing.com

www.TheInstitute.org

www.Presencing.com

www.TraumaHealing.com

About the Authors

Barry Lipscomb quietly soared his way from being a client of Michael Picucci's, to a mentee, and then to being a co-creator of the Focalizing technology.

Dr. Michael Picucci, PhD, MAC, SEP, brings decades of investigation and experience to his practice of Psychotherapy, Focalizing and Consulting. His professional expertise as a Psychologist, Licensed Psychotherapist, Master Addictions Counselor, Sexologist, Somatic Experiencing Practitioner and Organizational

Consultant has led to multiple awards for his contributions. In 2000, the National Institutes on Health (NIH), National Institute on Drug Abuse (NIDA), and the National Association of Professional Addiction Counselors (NAADAC) conjointly awarded him the "Outstanding Leadership in Research" honor for his many years of addiction/trauma fieldwork and reporting. At the award luncheon a senior NIH researcher engaged him on his research while offering a warning: "Michael, don't be disappointed if your findings are not mainstreamed for 20 years, if you are fortunate enough to create replicable models." Now seeing Focalizing as the required 'replicable model,' or 'human technology' Michael remembers those wise words. He continues his private practice while teaching and leading workshops around the globe. For further information visit Michael's website: www.Focalizing.com

Barry Lipscomb has been a life-long spiritual investigator and seeker. After brief use and study of Dr. Picucci's Focalizing discoveries, Barry compressed and reorganized Michael's findings from a more holistic perspective. From this grander view and their enjoyable, dynamic experience in planning workshops, they decided to write this book from within their Dynamic Link experience itself.

Barry also has over 20 years of experience as an operations executive and financial professional who has facilitated executive strategy sessions, led change initiatives, developed process improvements, mediated problem resolution and coordinated efforts across teams and between senior executives. He combines this experience with focalizing in order to tap into deepened awareness as a valuable business resource for creativity and innovation. Barry has synthesized a Conscious Business Operating Model, a process for Holistic Business Strategy, a system of accounting for the self-aware organization, and Management in the Moment, an agile management and project planning method. For more information visit www.BusinessFocalizing.com

CPSIA information can be obtained at www.ICGtesting.com
Printed in the USA
BVOW10s1256031113

335314BV00002B/3/P